Sandra Rodrigues Sarro Boarati

Maths and Technologies in Teaching Practice

Sandra Rodrigues Sarro Boarati

Maths and Technologies in Teaching Practice

Use of Information and Communication Technologies (ICT) by teachers at the Paula Souza Centre

ScienciaScripts

DEDICATORY

To God...

To my father Henrique Sarro (In Memoriam).

ACKNOWLEDGEMENT

To everyone who contributed to this project, especially my husband Artur and my children Rafaela and Henrique.

EPIGRAPH

"For the word of the Lord is true;

he is faithful in all that he does"
(Psalm 33:4).

ABSTRACT: This article sought to investigate through existing literature how information and communication technologies (ICT) are used as a form of production in the school environment and how those involved in the process make use of these tools in teaching practice. As it is a new tool, it is necessary to encourage older teachers to use these new means in order to achieve excellence, as the internet can help with guided school research, as well as the use of numerous software programmes to support teachers and students. In this way, a practical questionnaire was drawn up to help gather information on the use of ICT by teachers in the classroom and other relevant information. 61 teachers from different educational units of the Etecs of the Paula Souza Centre answered the questions. The answers provided by the teachers who took part in the survey made valuable contributions to understanding the position of these professionals when faced with the challenge of incorporating information and communication technologies into their teaching practice. This information was tabulated in graphical form to facilitate the study of the data, making it easier to carry out a critical analysis of the results and a thorough evaluation of the use of ICT in the classroom.

KEYWORDS: Mathematics; Technologies in Education; Information and Communication Technologies (ICT); Teaching Practice.

SUMMARY

CHAPTER 1

INTRODUCTION

Information and communication technologies (ICT) have brought many challenges to teachers, as the internet has provided unlimited access to information, bringing interaction and exchange of experiences between teachers and mutual exchange of teaching materials.

Little by little, this technology has made its way into homes, schools and has been integrated into schools as a new teaching methodology, integrating those involved and disseminating and sharing knowledge in a decentralised way.

The use of information and communication technologies (ICT) can be used in the classroom to facilitate and attract the interest of students, as a source of research, to work on interdisciplinarity, quality education, innovative and liberating education, authorship and autonomy on the part of students, etc (KESSLER, 2015, p. 11).

There are many challenges to integrating ICT into a school environment and educational activities. They are of great importance for improving the quality of teaching and learning, providing countless benefits for those who know how to use them.

As it is a new tool, it is necessary to encourage older teachers to use these new means in order to achieve excellence, as the Internet can help with guided school research as well as the use of numerous software programmes to support teachers and students.

There are many difficulties in preparing more dynamic lessons due to the lack of time, teacher resources and also student behaviour and commitment. Students, on the other hand, prefer dynamic lessons because they feel more motivated and learn better (FELIX, 2014, p. 6).

In the process of implementing technology at school to complement the student's education, the teacher was not part of it, so it is necessary to quickly "keep up with social transformation and the speed with which information is released in the media, so that the student can acquire meaningful knowledge from contextualised information" (CANTINI et al, 2006, p.880).

There is still a barrier to changing classroom attitudes towards ICT, as it requires teachers to update their teaching practices and students to be open to new ways of learning the content.

According to Pontes (2000, p. 64), Information and Communication Technologies (ICT) "represent a determining force in the process of social change, emerging as the backbone of a new type of society, the information society".

Information is an instrument that promotes socialisation and interaction between people. It is clear that there is a need for people to undergo a process of adaptation in relation to information and communication technologies (ICT), because it changes and transforms the vision of teachers and students in a school environment, bringing facilities and improvements in the presentation of programme content.

1.1 Justification

The classroom is built with the aim of making it a place of meaningful learning with the interaction of teacher and student. In order for this interaction to be efficient, changes are needed in relation to working techniques - blackboard, chalk and something else. With the natural evolution of society, now known as the "Information Society", it has become necessary to include technology in classroom practices, such as radio, television, videos, DVDs, data shows, computers and the internet. These are the new ICT (Information and Communication Technologies) tools that are of great importance for improving the quality of teaching and learning, bringing benefits to the school environment. However, it is extremely necessary to know how to use these tools for the benefit of education.

This study seeks to reveal how these technological resources can be an ally in building a new relationship between students and knowledge and to what extent they can increase teaching practice in five Paula Souza Centre schools. In order to see how these powerful tools can help teachers and also to assess whether their resources are used in the classroom, a questionnaire was applied to evaluate these relationships between learning, the student, the teacher, the school and technology.

1. 2General Objective

The general objective of this work will be to identify and show the importance of Information and Communication Technologies (ICT) in the teaching-learning process in teaching practice and the level of involvement of effective education professionals that leads to the quality of the teaching-learning process through technological resources.

1. 3Specific Objectives

The specific objectives of this work will be:

-Investigate   the impact of ICT in the classroom;

-To find out   how often teachers use this resource;

-       Find out which are the main technological tools used;

-Identify   the ICT infrastructure of schools;

-To find   out the reasons for using ICT.

1. 4 Methodology

The work will consist of research into existing literature to gather data on the use of Information and Communication Technologies (ICT) in teaching practice.

Qualitative research will be carried out using an electronic questionnaire with closed questions applied to primary and secondary school teachers at the Etecs of the Paula Souza Centre. The questionnaire will cover

-Whether the teacher uses ICT at school;

-Frequency of   use;

-The reason   you use them;

-Whether this practice is well received by the students;

-Which   tools teachers   use   the   most (computer, data show, internet, TV, DVD, etc.);

-Whether the school has adequate infrastructure for the use of technological resources;

-Whether there are any specific technological tools for education in the school.

CHAPTER 2

INFORMATION AND COMMUNICATION TECHNOLOGIES (TIC)

Much has been said about the inclusion of technological resources in the school environment. The discussion has intensified in recent decades with the popularisation of the personal computer, smartphones, mobile phones and broadband internet.

According to Rodrigues (2009), teachers receive many demands to incorporate information and communication technologies (ICT) into their classrooms, as official educational documents recommend their use:

"There is an indisputable growing need for students to use computers as a school learning tool, so that they can keep up to date with new information technologies and be equipped for present and future social demands" (BRASIL, 1998, p. 96).

"Communication and information technologies and their study should permeate the curriculum and its subjects" (BRASIL, 1999, p. 134).

According to Rodrigues (2009, p. 2), these documents are guidelines that "exert a certain influence on teaching, but it is from the day-to-day relationship with students that the demand for diversifying resources and making the most of their possibilities in educational activities comes".

The issues involved revolve around the benefits of information and communication technologies (ICT) as an augmentation to teaching methodology and the extent to which they can make the classroom more attractive, as well as recognising how technology, at various times, is responsible for access to new forms of knowledge.

On the other hand, it must be admitted that, for many educators, this virtual and technological environment is still little explored and difficult to adapt to teaching practice.

According to Pontes (2000, p. 64):

> We currently find very different attitudes among teachers towards information and communication technologies (ICT). Some view them with suspicion, trying to postpone the moment of the unwanted encounter as long as possible. Others use them in their daily lives, but are not sure how to integrate them into their professional practice. Still others try to use them in their classes without changing their practices. An enthusiastic minority are forging ahead, endlessly exploring new products and ideas, but are faced with many difficulties as well as perplexities.

We realise that we need to mitigate this impact in the classroom and use the talent that comes from information and communication technologies (ICT) to show that there are new ways of passing on knowledge to students.

2.1 Types   of ICT

ICTs are available for use at any time.

According to Época magazine (20/06/2011) we have a timeline for ICTs:

In 1500, the so-called booklet was made available to teachers, which was later passed on to students as a source of reference and activities, as shown in Figure 1.

Figure 1 - Booklet (Época Magazine, 2011)

As early as 1900, the blackboard was a striking instrument and still prevails today with some innovations using pen instead of chalk, as shown in Figure 2.

Figure 2 - Blackboard (Época Magazine, 2011)

The pencil, which is still an important tool today, appeared in 1900, as shown in Figure 3.

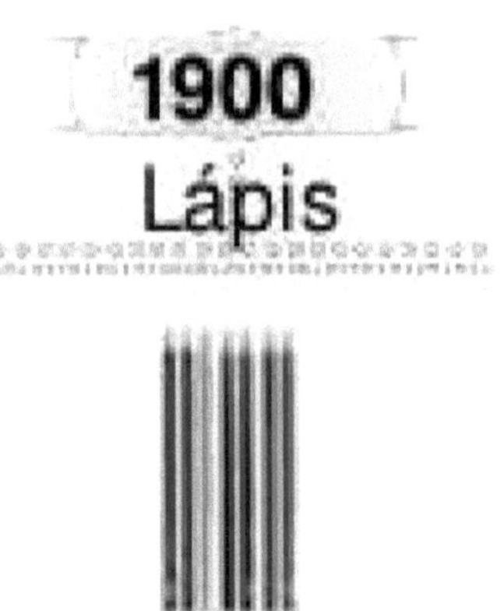

Figure 3 - Pencil (Época Magazine, 2011)

The slide projector (photo) appeared in 1950 and was very useful for visualising photos and documentaries, as shown in Figure 4.

Figure 4 - Slide Projector (Época Magazine, 2011)

In 1960, the mimeograph appeared to help teachers copy tests, activities and other types of documents, as shown in Figure 5.

Figure 5 - Mimeograph (Época Magazine, 2011)

The Data Show appeared in 1990 and replaced the slide projector, as shown in Figure 6.

Figure 6 - Data Show (Época Magazine, 2011)

At the same time, 1990, the computer also appeared, as shown in Figure 7.

Figure 7 - Computer (Época Magazine, 2011)

Little by little, computer models were revamped and made to look lighter. In 2002, the Laptop appeared, as shown in Figure 8.

Figure 8 - Laptop (Época Magazine, 2011)

And finally, in 2011, the tablet, which is widely used in schools for teaching purposes, appeared, as shown in Figure 9.

Figure 9 - Tablet (Época Magazine, 2011)

Nowadays, old mobile phones have become mini-computers with the advance of technology. They now have Android and Apple processors to access the internet and applications.

What we see being used most in schools today are tablets, mobile phones, notebooks and data shows.

CHAPTER 3

DISCUSSION: TECHNOLOGICAL RESOURCES

The innovation of a society that is rapidly reorganised by new technologies places new demands on teachers, requiring professionals capable of building knowledge quickly and innovatively (REHEM, 2016).

According to Rehem (2016, p. 7):

> The need for teachers to undertake their work, act with autonomy and responsibility, be willing to take initiatives, solve problems creatively, have systemic thinking and good communication skills in various languages and technologies is consistent with the demand for new skills from the teachers who train these professionals. This reality points to the need for a critical awakening regarding the profile of this teacher required to train new professionals. This profile for the new teacher points to the need for renewed teacher training, which also provides transformative intellectuals with the opportunity to build the competences and knowledge that underpin their formative work. Although complex, the challenge is stimulating to promote innovation in this field.

The resources used at school, such as blackboards, chalk, textbooks, pencils, oral explanations and the use of language "are part of educational technology, along with TV, overhead projectors, video and computers" (CARNEIRO, 2002, p. 49).

According to Seeger et al (2012, p. 1890), the school must present itself as "an environment capable of immersing these technologies in the service of a teaching methodology that favours the interaction of students in this information society, thus cancelling out the social differences that are not relevant to this process".

Bringing the school closer to the advances made by society in terms of the transformation, production and transmission of information favours a reduction in the gap between the school environment and the student's life - which would also reduce "[...] the differences in opportunity between public and public schools, which are increasingly computerised" and points out that "[...] there is little discussion of which modes of computerisation are being worked on and for what purpose" (CARNEIRO, 2002, p. 50-51).

The world is constantly evolving, due to technological and scientific advances that have provided humanity with progress in all areas of knowledge. Education is no different; it has evolved with the emergence of methodologies, theories and the inclusion of ICT in the teaching-learning process. With the context promoted by ICT and its constant development comes the need to make the most of technological resources as tools in the teaching-learning process.

There is a need to rethink teaching practice, because we cannot follow the model of the last century characterised by a conservative, repetitive and teacher-centred pedagogical practice. Currently, the learning relationship through ICT has articulated the relationship between teachers and students to inform, communicate, interact and learn, constituting a language of synthesis, encompassing aspects of orality and writing in new concepts.

It will always be a challenge to review one's own educational practices in order to enrich them, "whenever possible, new knowledge for teachers who will be investigating and reflecting on their teaching actions, thus seeking teaching strategies so that the student appropriates the elaborated knowledge in a meaningful way" (CARVALHO, 2005, p. 2).

Many teachers believe that "they will not learn and will have difficulties using ICT due to lack of knowledge, insecurity, traditionalism, resistance or lack of adequate training; lack of time to plan lessons using such resources" (SANTOS, 2016, p. 119).

It should be emphasised that technologies and methodologies "incorporated into teaching knowledge change the traditional role of the teacher, who sees throughout the educational process that his or her pedagogical practice needs to be constantly reassessed" (PEREIRA, 2012, p. 13).

Regardless of the use of ICT, teachers must have solid and efficient academic training, with methods that teach children and adolescents to be more humble and humane, making them understand aspects that involve racial, ethnic and ethical problems, among others.

In addition, teachers need to redefine their role and their interaction with students and current innovations. The teacher is considered an important factor in ensuring the integration of new technologies into the school curriculum.

Therefore, their training in this area should be the object of special attention, as it is a conditioning factor for the successful implementation of technological resources, the computer in particular, as a teaching tool.

It is well known that students are in direct contact with technology. Knowing this, schools and teachers need to "propose methodologies that use new information and communication technologies to achieve significant results, experiencing affective interpersonal and group communication and participation processes" (ESPÍRITO SANTO, 2012, p. 1023).

The use of information technology in education is already widespread. The technical schools of the Paula Souza Centre have their own classrooms for using computers (computer labs) and, regardless of the course chosen, there is a specific curricular component for this area of knowledge. In addition, several classrooms are equipped with data shows and other laboratories with specific features that meet the basic learning needs for certain courses.

But the discussion that surrounds this work is how we can use these resources more productively and whether teachers can take advantage of this immense digital universe.

3.1 Applicability

In order to obtain the opinion of education professionals who are in the classroom, specifically in five Etecs of the Paula Souza Centre, an electronic questionnaire was applied to teachers to obtain data on teaching practice using information and communication technologies (ICT), consisting of ten closed questions.

The decision to send the questionnaire electronically was a proposal that "shows immediately, through the time and quantity of returns of the completed instrument" (PORTO ALEGRE, 2005, p. 143).

With 61 questions tabulated, the focus of the analysis was on the teachers' position in the face of the new challenge, i.e. what impact this technology has on the academic environment. From the answers obtained, it can be said that:

1.   When asked about the need for a new pedagogy in relation to the evolution of information: 86.9 per cent of the teachers interviewed said that they believe that the evolution of information requires a new pedagogy; 9.8 per cent believe that perhaps a new technology is needed and 3.3 per cent do not believe so, according to Figure 10.

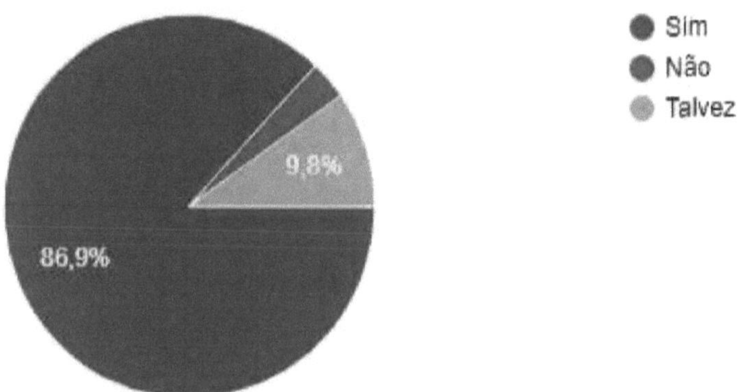

Figure 10. The evolution of information requires a new pedagogy

2.   Regarding the way in which ICT has a positive impact on relations between pupils and teachers: 83.3 per cent of the teachers interviewed said that they believe that ICT has a positive impact on relations between pupils and teachers; 13.3 per cent said that it might have a positive impact and 3.3 per cent did not believe so, according to Figure 11.

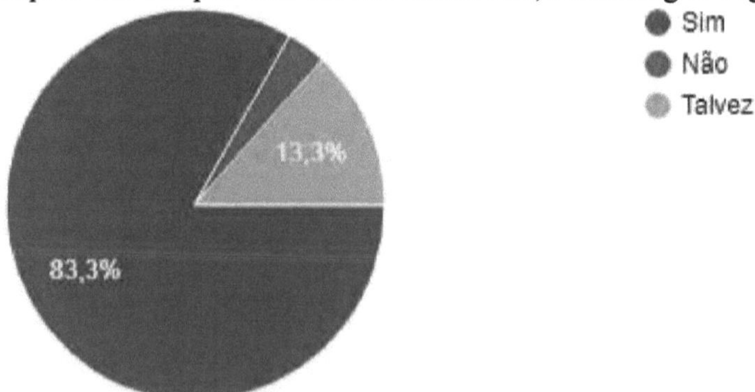

Figure 11. ICT positively changes relationships between students and teachers

3.   Another question asked was about the transformative power of

ICT in the school environment: 83.3 per cent of the teachers interviewed said that they believed ICT to be a transformative force in the school; 10 per cent said that it might be a transformative force and 6.7 per cent did not believe so, according to Figure 12.

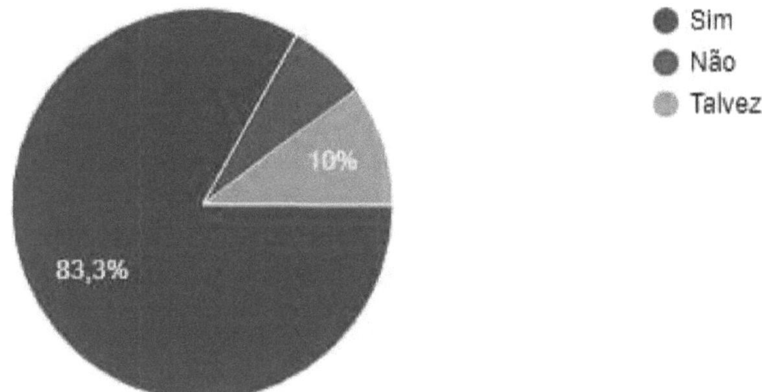

Figure 12. ICT is a transformative force at school

4.   The questionnaire also revealed how often they use these resources and which tools they use the most: 64.4% of the teachers interviewed said that they use ICT frequently; 32.2% use it sometimes and 3.4% use it rarely, as shown in Figure 13.

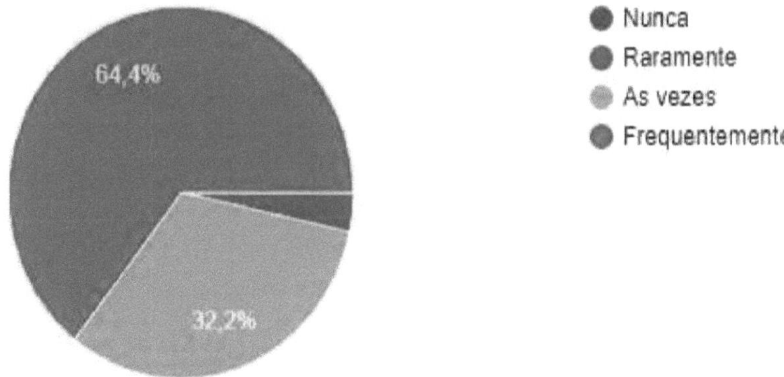

Figure 13. Frequency of use

5   Teachers were asked why they use these tools: 93.4% of teachers

use the Data Show; 82% use the computer; 75.4% use the internet; 49.2% use TV/DVD; 29.5% use the AVA environment; 6.6% use tablets; 6.6 use a camcorder; 4.9% use a radio and 3.3% use a tape recorder, according to Figure 14.

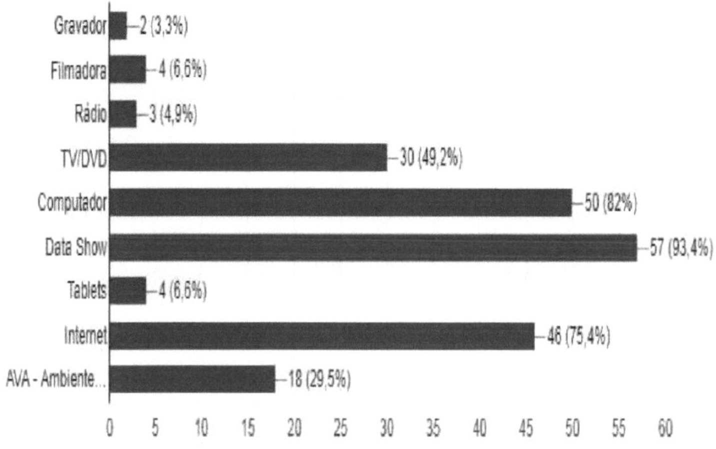

Figure 14. Main tools

6.   The teachers were asked why they use these tools and they were presented as follows: 57.4% of the teachers interviewed said that they use the resources to provide dynamic lessons; 24.6% use them because of the natural evolution of the work environment; 8.2% use them to stimulate curiosity; 6.6% use them to increase motivation and 3.3% use them to promote integration between students, as shown in Figure 15.

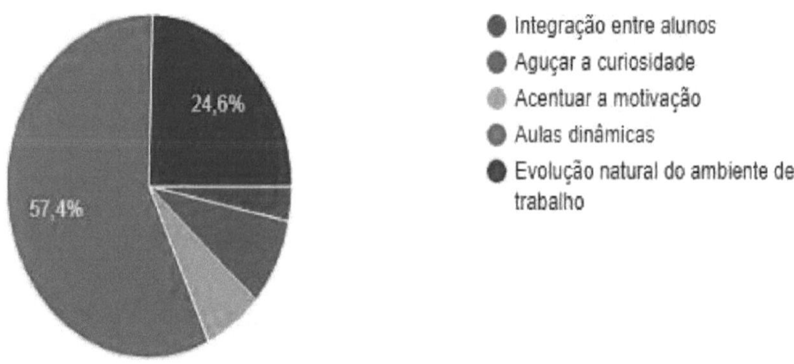

Figure 15. Purpose of utilisation

7.   Another point raised was student learning in relation to the use of these resources: 60.7 per cent of the teachers interviewed said that in relation to student learning they rated the use of these resources as good; 36.1 per cent rated it as excellent and 3.3 per cent rated it as fair, as shown in Figure 16.

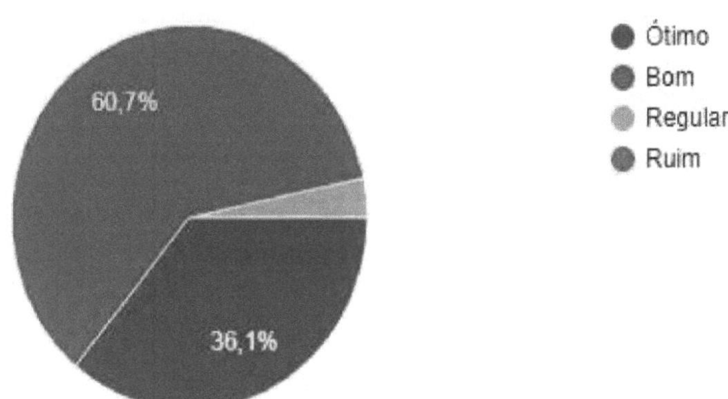

Figure 16. Student learning

8.   Another positive finding was the technological tools the school has: 70 per cent of the teachers interviewed said that their school has some specific tool for education and 30 per cent said that it doesn't, as shown in Figure 17.

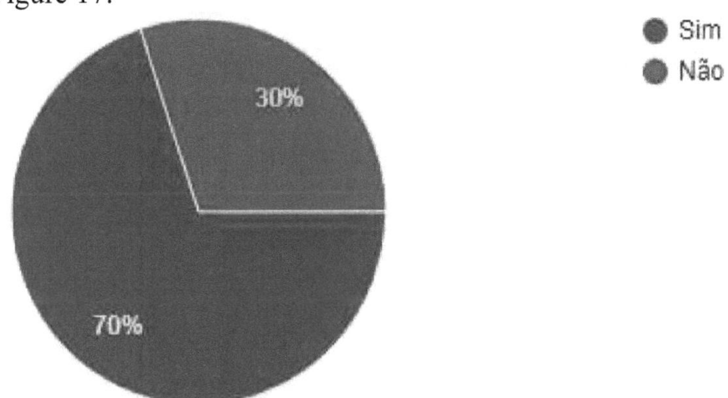

Figure 17. Student learning

9.   The methodology also sought to ascertain teachers' feelings of

security when using ICT: 86.9 per cent of the teachers interviewed said that they feel secure or comfortable using technological resources; 11.5 per cent said that they sometimes don't feel secure or comfortable and 1.6 per cent said that they don't feel secure or comfortable using the resources, as shown in Figure 18.

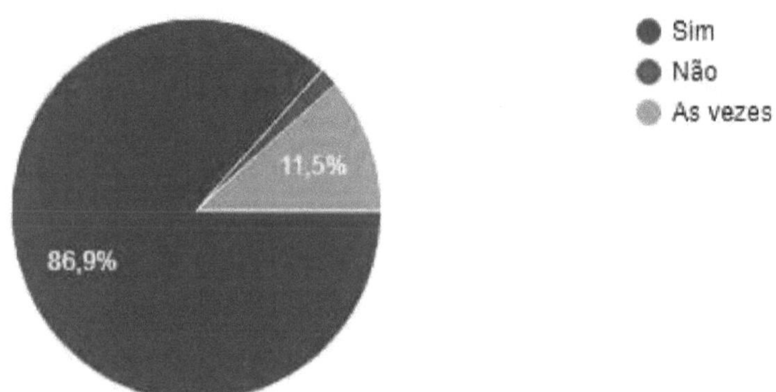

Figure 18. Safety or comfort when using technological resources

10.   Another question about school infrastructure was asked in the questionnaire: 82 per cent of the teachers interviewed said that the school has adequate infrastructure for using technological resources and 18 per cent said that sometimes they don't feel safe or comfortable, as shown in Figure 19.

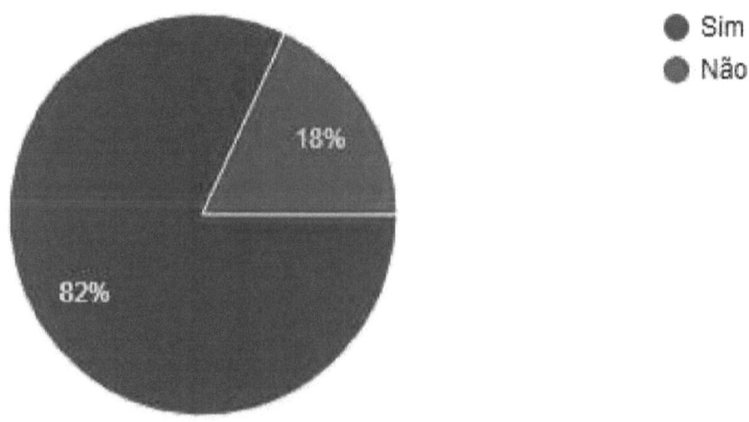

Figure 19. Adequate infrastructure

CHAPTER 4

WORK DONE IN CLASS

It can be seen that by applying interdisciplinarity between disciplines, such as mathematics and biology, the conceptualisation and practical application of statistical concepts was of great value in obtaining the results. 120 Students from the Etec of the Paula Souza Centre from the 1st Etim of Automation, Informatics and Logistics in the subject of Mathematics prepared the activity displayed inside the school, as shown in Figures 20 to 22.

Figure 20 - Flowers/leaves
Source: Author (2017)

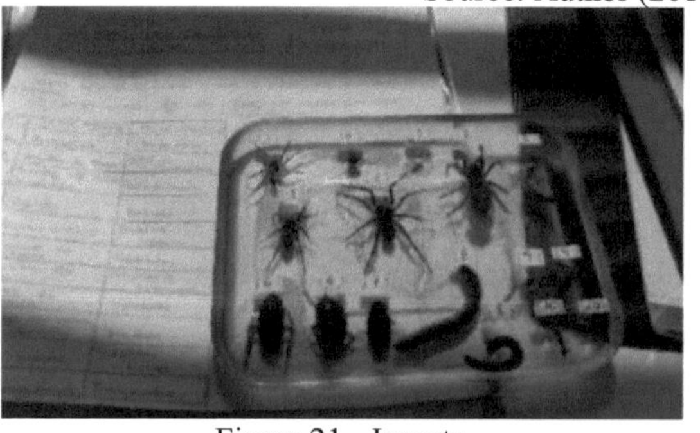

Figure 21 - Insects
Source: Author (2017)

Figure 22 - Stones

Source: Author (2017)

The application of statistics concepts related to the environment (Biology) enabled the students to tabulate the data according to the chosen population sample, providing the mode, median, arithmetic mean, absolute frequency and relative frequency.

All the practical activities were exhibited in the school after completion and released for appreciation for 15 days, after which the games were sent to the library to be available for students to play in class breaks.When working with Cartesian planes in maths, you can go beyond just expressions and equations in graphs. In Figures 23 to 25, the students applied the construction of symmetry in maths in a joyful and relaxed way.

Figure 23. Wind symmetry

Source: Author (2017)

Figure 24. Butterfly symmetry

Source: Author (2017)

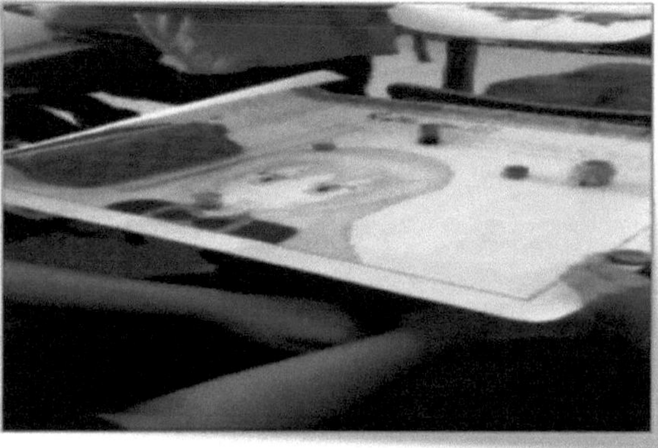

Figure 25. Face symmetry
Source: Author (2017)

In order to build symmetry, it was necessary to identify which type of symmetry to use, taking into account the work with angles so that the symmetry was well done. Mathematical games also stimulate students to learn and play with maths. The games promote interaction between the students, because as well as evaluating the games they were able to play and compete with each other and with the class, as shown in Figures 26 to 29.

Figure 26. Maths games
Source: Author (2017)

Various Games

Source: Author (2017)

Figura 27.   Board game
Source: Author (2017)

Figure 29. Naval battle
Source: Author (2017)

The games were made by the students themselves, describing the game step-by-step, the method, the rules and the construction of the game itself.

Once the games were finished, all the students swapped the games between them and played, testing the rules and advancing the squares according to the correctness of the mathematical expressions and backtracking if any of the answers were wrong.

All the practical activities were exhibited in the school and released for appreciation for 15 days, after which the games were sent to the library to be available for students to play in between classes.

Information and communication technologies (ICT) have brought many challenges to teachers, as the internet has provided unlimited access to information, bringing interaction and exchange of experiences between teachers and mutual exchange of teaching materials.

4.1 Use of Information and Communication Technology (ICT)

Use of Information and Communication Technology (ICT) in the classroom for teaching purposes: To show the importance of ICT in the classroom for teaching purposes, as an example we will have the mobile phone which is a mini-computer used for applications aimed at teaching maths, dictionaries, office suite, etc. Information and communication technologies (ICT) have brought many challenges to teachers, as the internet has provided unlimited access to information, bringing interaction and exchange of experiences between teachers and mutual exchange of teaching materials.

Little by little, this technology has made its way into homes and schools and has been integrated into schools as a new teaching methodology, integrating those involved and disseminating and sharing knowledge in a decentralised way.

The use of information and communication technologies (ICT) can be used in the classroom to facilitate and attract the interest of students, as a source of research, to work on interdisciplinarity, quality education, innovative and liberating education, authorship and autonomy on the part of students, etc.

There are many difficulties in preparing more dynamic lessons due to lack of time, teacher resources and student behaviour and commitment. Students, on the other hand, prefer dynamic lessons because they feel more motivated and learn better (FELIX, 2014, p. 6).

For the use of maths, ICTs have come to add value, because through these technologies you can use software to help with maths, such as Excel, Geogebra, etc.

According to Rodrigues (2009), teachers receive many demands to incorporate information and communication technologies (ICT) into their classrooms, as the official education documents themselves state.120 Students from the 1st year of the Automation, Informatics and Logistics course in Maths prepared the activity displayed inside the school, as shown in Figures 30 to 32.

Figure 30 - ICTs
Source: Author (2017)

Figure 31 - ICT
Source: Author (2017)

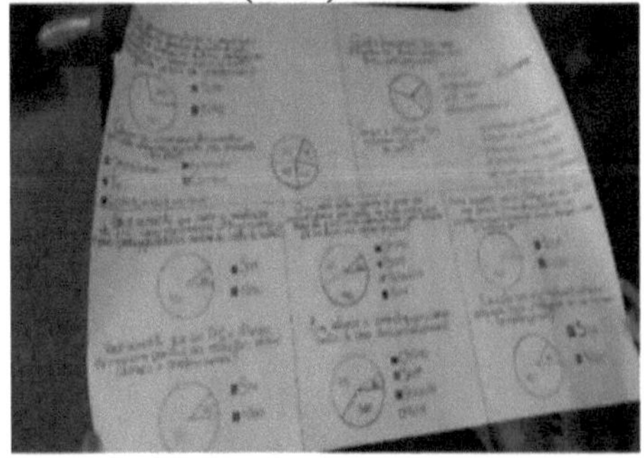

Figure 32 - Mobile phone use
Source: Author (2017)

For the content of the 3rd year, more geometric figures are studied and because it is linked to angles, sine, cosine and tangent, the students already have a certain rejection for this part of maths because they don't master the content.

As a solution to this type of situation, there are various pieces of software available that can be applied to maths. One widely publicised example is the construction of solids physically and using Geogebra software. Figures 33 and 34 were constructed using Geogebra.

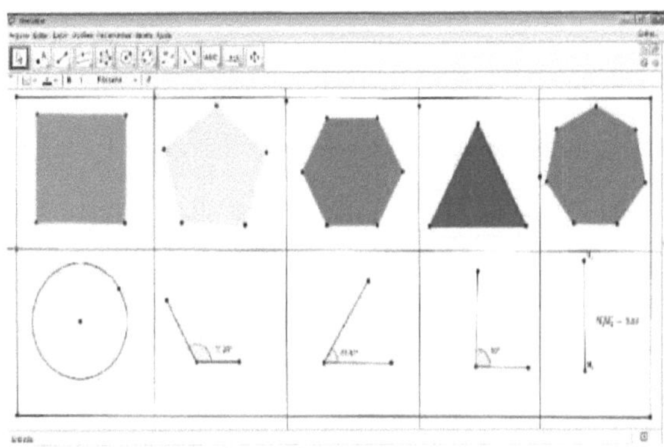

Figure 33. Geogebra - Geometry
Source: Author (2017)

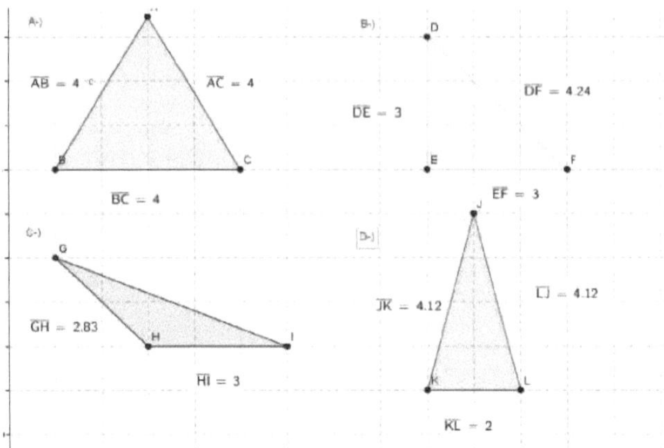

Figure 34. Geogebra - Angles
Source: Author (2017)

4.2 Application of Seminars

The application of the seminars will include:

- The types of information system models used in Logistics: ERP (Enterprise Resource Planning) and MRP (Materials Requirements Planning).

In order to carry out this activity, it was necessary to look at the competences for applying IT in Logistics: defining information systems according to the needs and limitations of the organisational structure; assessing the hardware and software needed to control and monitor the organisation's operational activities. The skills include: differentiating the various forms of organisation of different types of company; identifying computerised systems for recording and monitoring corporate processes; using corporate programmes and systems to record and monitor established targets and controls; collecting information to monitor the activities of all the company's sectors; selecting new technologies in the logistics area;

running applications to help with decision-making in the logistics area. he work will consist of research into existing literature to gather data on the use of Information and Communication Technologies (ICT) in teaching practice.

Another way of working with diversity in the classroom is to use seminars, which are group assignments.

The seminar is considered a technique, but also an effective method of study, so it is a didactic-pedagogical practice usually applied in groups using the genre of oral exposition.

According to Campos (2006, p.8), the seminar is a "form of group work widely used in secondary, post-secondary, undergraduate and postgraduate education as a socialising teaching technique".

For Costa and Baltar (2009), this pedagogical practice is a language action that provokes the exercise of criticism, of defending one's point of view on something, thus developing students' discursive competence, both orally and in writing.

An example of application was the seminar on ERP, a system widely used in logistics.

ERP stands for Enterprise Resource Planning. It is a computer system responsible for taking care of all the daily operations of a company. It helps the company to be more organised and increases its competitiveness in the market (GASPAR, 2012). For the ERP systems available on the market, it is possible to find all this information within a single system, as shown in Figures 35 to 37.

Figure 35 - Logistics students
Source: Author (2017)

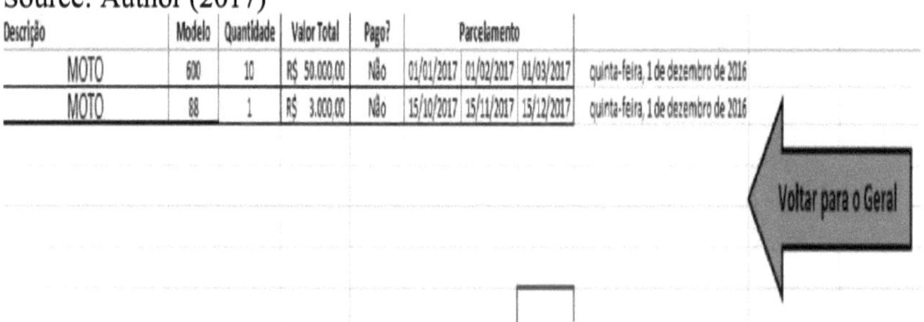

Figure 36 - ERP system
Source: Author (2017)

Descrição	Modelo	Quantidade	Valor Total	Pago?	Parcelamento			
MOTO	600	10	R$ 50.000,00	Não	01/01/2017	01/02/2017	01/03/2017	quinta-feira, 1 de dezembro de 2016
MOTO	88	1	R$ 3.000,00	Não	15/10/2017	15/11/2017	15/12/2017	quinta-feira, 1 de dezembro de 2016

Figure 37 - ERP system

Source: Author (2017)

For Paz, Nascimento and Silva (2016, p. 1), the use of seminars is a "pedagogical tool that influences the teaching-learning process, as well as the composition of the disciplinary assessment carried out by teachers".

All these tools used to improve maths lessons and other subjects promote more practical and efficient learning.

CHAPTER 5

CONCLUSION

The answers provided by the teachers who took part in the research made valuable contributions to understanding the position of these professionals when faced with the challenge of incorporating information and communication technologies into their teaching practice.

Some obstacles to implementing these resources were perceived, but the survey revealed that, even so, most of the interviewees admit that technology has created a new school environment. It also showed that these tools are welcome and are being used more and more, with greater frequency and quality.

Another important reflection in the article reveals that this new scenario allows for new exchanges of knowledge, new forms of integration and new types of behaviour. ICT has brought a new language into the classroom and that's why it's so important to create ways to emphasise its resources.

In the theories formulated by Vygotsky, it can be seen that his precepts fit in very well with the association between learning and ICT. By pointing out that the individual internalises certain ways of functioning given by culture and transforms them into instruments of thought and action, Vygotsky laid the foundations for a new understanding of the relationship between man and the historical context, placing him in the complex web of social relations. In this sense, with the natural evolution of technology, the insertion of these technological elements into the pedagogical process is irreversible, as it is already moulded into the contemporary cultural and social transformation. This transformation of language is even more accentuated in young people who were born into a "digital world" (MEIRA, 1998).

Finally, it should be borne in mind that in carrying out his or her teaching role, the teacher is not only teaching certain content, but also forming individuals through interpersonal relationships that express a whole network of social values that are not always fully realised. At the same time, it is also a process of preparing students for the development of these relationships (inside and outside school). Technological resources can be used to enhance these relationships (social, cultural and learning). However, as educators, we need to be aware of these transformations in order to take advantage of them and apply them in the best way at school.

CHAPTER 6

BIBLIOGRAPHICAL REFERENCES

BRAZIL. Secretariat of Basic Education. Parâmetros Curriculares Nacionais: terceiro e quarto ciclos do ensino fundamental: introdução aos parâmetros curriculares nacionais. Brasília: MEC/SEF, 1998.

_____ . Ministry of Education, Secretariat of Secondary and Technological Education. National Curriculum Parameters: Secondary Education. Brasilia: Ministry of Education, 1999.

CANTINI, Marcos Cesar; BORTOLOZZO, Ana Rita Serenato; FARIA, Daniel da Silva; FABRÍCIO, Fernanda Biazetto Vilar; BASZTABIN, Rogério; MATOS, Elizete. The teacher's challenge in the face of new technologies. Available at: <http://www.pucpr.br/ eventos/educere/educere2006/anaisEvento/docs/CI- 081-TC.pdf>. Accessed on: 23 June 2017.

CARNEIRO, Raquel. Informática na educação: representações sociais do cotidiano. 2. ed. São Paulo: Cortez, 2002.

CARVALHO, Rosiani. Technologies in everyday school life: possibilities for articulating pedagogical work with technological resources. 2005. Available at: < http://www.diaadiaeducacao.pr.gov.br/portals/pde/arquivos/1442- 8.pdf>.

Accessed on: 23 Jun 2017.

ESPÍRITO SANTO, Janete Araci do; CASTELANO, Karine Lôbo; ANDRÉ, Bianka Pires. The use of technology in teaching practice: a case study in the context of a public school in the interior of Rio de Janeiro. II International ICT and Education Congress . 2012. Available at:

<http://revistas.utfpr.edu.br/pb/index.php/revedutec-ct/article/view/1554>. Accessed on: 23 Jun 2017.

FELIX, Camila Danielle de Souza. The influence of methodologies on *science* teaching in *a public school in the municipality of Rancho Aleg.e D'Oeste, Paraná.* 2014Available in : <http://repositorio.roca.utfpr.edu.br/jspui/handle/1/4310>. Accessed on: 23 Jun 2017.

KESSLER, Maria de Lourdes Moraes. *The use of technology in teaching practice.* UFRGS. 2015. Available at : <http://www.lume.ufrgs.br/handle/10183/133912>. Accessed on: 23 Jun 2017.

MEIRA, Marisa Eugênia Melillo. Development and learning: reflections on their relationship and implications for teaching practice. *Ciência & Educação,* 1998 ,v.5 ,n.2, p.61-70. Available at: <http://dx.doi.org/10.1590/S1516-73131998000200006>. Accessed on: 08

Aug 2017.

PEREIRA, Bernadete Terezinha. *The use of information and communication technologies in school pedagogical practice*. 2012. Available at: <http://www.diaadiaeducacao.pr.gov.br/portals/pde/arquivos/1381-8.pdf>. Accessed on: 23 Jun 2017.

PONTE, João Pedro da. Information and communication technologies in teacher training: what challenges? *TIC na educação*, n. 24, Sep-Dec 2000. Available at: <http://rieoei.org/rie24a03.htm>. Accessed on: 20 July 2017.

PORTO ALEGRE, Laíze Márcia. The use of information and communication technologies in teaching practice at a technological teaching institution. 2005. 225f. Thesis (Doctorate in Education) - State University of Campinas, Campinas, 2005. Available at:< http://repositorio.unicamp.br/jspui/handle/REPOSIP/253081>. Accessed on: 23 Jun 2017.

REHEM, Cleunice. The vocational education teacher: what profile corresponds to contemporary challenges? Senac Technical Bulletin, 25 August 2016. Available at: < http://www.senac.br/BTS/311/boltec311d.htm 1/10>. Accessed on: 20 July 2017.

RODRIGUES, Nara Caetano. Information and communication technologies in education: a challenge in teaching practice. *Fórum Linguístico.* Florianópolis, v.6,   n.1, p. 1-22, jan-jun, 2009. Available at: <https://www.faecpr.edu.br/universidadevirtual/artigos/artigo_tecnologia_da_informacao_e_comunicacao_na_educacao.pdf>. Accessed on: 23 June 2017.

SANTOS, Domingas Cantanhede dos. *Information and communication technologies in*

teaching pedagogical practice. 2016. 149f. Dissertation (Master's in teaching) - Univates University Centre, Lajeado, 2016. Available at:<https://www.univates.br/bdu/bitstream/10737/1047/1/2016DomingasC an tanhededosSantos.pdf>. Accessed on: 23 June 2017.

SEEGGER, Vania; CANES, Suzy Elisabeth; GARCIA, Carlos Alberto Xavier. Technological strategies in pedagogical practice. UFSM, Pampa, 2012.   Available   at:< https://periodicos.ufsm.br/remoa/article/viewFile/6196/3695>.   Accessed on: 23 Jun 2017.

yes I want morebooks!

Buy your books fast and straightforward online - at one of world's fastest growing online book stores! Environmentally sound due to Print-on-Demand technologies.

Buy your books online at
www.morebooks.shop

Kaufen Sie Ihre Bücher schnell und unkompliziert online – auf einer der am schnellsten wachsenden Buchhandelsplattformen weltweit! Dank Print-On-Demand umwelt- und ressourcenschonend produziert.

Bücher schneller online kaufen
www.morebooks.shop